Tropical Blossoms of Costa Rica

Text copyright © 2008 Marc Roegiers and John K. McCuen
Photographs copyright © Turid Forsyth

Cover photograph of *Trichopilia suavis* by Turid Forsyth.
Back cover photographs of *Hamelia patens* (top), *Hedychium coronarium* (middle), and *Ageratum* (bottom) by Turid Forsyth.

The photographs on p. 9 and p. 23 were contributed
by William A. Haber and are reproduced by permission.

The photographs on p. 39 and p. 54 were contributed
by Adrian Hepworth and are reproduced by permission.

ISBN: 978-0-9705678-8-8

Printed in China

10 9 8 7 6 5 4 3 2 1

Developmental editor: David Featherstone
Book design: Ecce Gráfica; Zona Creativa, S.A.
Ecce Gráfica designer: Eugenia Picado
Zona Creativa designer: Gabriela Wattson

Published by Distribuidores de la Zona Tropical, S.A.
www.zonatropical.net

Tropical Blossoms
of Costa Rica

Willow Zuchowski

Photographs by Turid Forsyth

A Zona Tropical Publication

Introduction

The publication of both this book and its companion volume, *Tropical Trees of Costa Rica*, follows closely on the heels of the author's substantially longer and more technical *Tropical Plants of Costa Rica: A Guide to Native and Exotic Flora*. That first book was received with such enthusiasm—by botanists and horticulturalists alike—that the publisher was eager to produce a series of more compact books for the popular audience.

As you glance through the many beautiful photographs included here, you'll note that this book carries a somewhat imprecise title; a number of the plants, in fact, bear foliage or fruits that are more striking than their blossoms. In one way or another, then, nearly every species in *Tropical Blossoms of Costa Rica* is striking; this is, however, more than just a catalog of beautiful plants. How is a specific plant pollinated? What's the origin of its common name/s (its scientific name)? Does it have any medicinal properties—either real or apocryphal? Has it played an important role in the life of indigenous cultures? Does the plant produce edible fruit? Can you grow it in your garden? By answering these and other questions—if only for a small subset of all the plants in Costa Rica—the author hopes to provide the reader with some sense of the diversity of flora in Costa Rica; the complexity of the relationships that exist between plant and animal; and the importance of preserving the country's natural environments.

In this book, plant descriptions appear in alphabetical order—arranged by English common name. Consult the general index, page 79, to find a list of both scientific names and English and Spanish common names. If you know what a plant looks like—but you don't know its name—consult the visual index, beginning on page 75, to quickly find out if it is included in this book.

Note: Measurements are expressed in both metric and nonmetric terms. The metric measurements that are cited here come from the author's previous book, *Tropical Plants of Costa Rica*. When converting from centimeters to inches—and meters to feet—the editor rounded off numbers; in the case of converting from centimeters to inches, the resulting discrepancy is trifling.

Left: *Epidendrum radicans* orchid.

Ageratum, Santa Lucía
Ageratum species

Family: Asteraceae

The genus *Ageratum* contains around forty-four species, with at least seven species found in Costa Rica. In Spanish, the various species are known as *Santa Lucía* and are associated with early dry season, when fields become covered with the lavender-blue flowers. A Costa Rican custom is to give someone a little bouquet of these flowers in January to wish them a prosperous year. The males of certain clear-wing butterflies take nectar from the tiny florets of some *Ageratum* species, stocking up on alkaloid chemicals that serve as building blocks for the pheromones they use to attract females. During mating, the females may receive some of these alkaloids, which also protect the butterflies from predators by making them taste bad. Ageratums can grow to 3 ft (1 m) tall, but most often they are shorter; the leaves are usually opposite. They have bluish flower heads, and what appears to be one flower is actually many tiny ones packed together. The genus *Ageratum* is found from Mexico to South America, as well as in the West Indies. It belongs to one of the largest and most diverse families of flowering plants, the aster family, which includes daisies, sunflowers, goldenrods, and artichokes. The genus is widespread in Costa Rica, occurring along forest edges and in pastures, in full or partial sun, from 165 to 8,200 ft (50 to 2,500 m) or higher, with several species growing mostly in the mountains.

Arum lily, Cala
Zantedeschia aethiopica

Family: Araceae

The arum lily grows in clumps under 3 ft (1 m) tall. It has arrow-shaped leaves, 8 to 16 in (20 to 40 cm) long. A white funnel-like spathe about 6 in (15 cm) tall envelops a yellow-orange spadix with female flowers on the lower portion and male flowers above. The arum lily, which is also called calla lily, *cala*, and *cartucho* (which means paper cone in Spanish), is originally from South Africa; in Costa Rica, it is naturalized in damp or wet areas, from mid to high elevations. It is common in spots on the route over Cerro de la Muerte and toward the high volcanoes Poás and Irazú, where these ornamentals have escaped into moist, cool pastures. In Costa Rica, as elsewhere, this is a popular choice for cut-flower arrangements for funerals. Many artists, including Henri Matisse, Georgia O'Keeffe, and Diego Rivera, have chosen arum lilies as subjects for their paintings. Although some sources list this as a poisonous plant, people use it topically to soothe burns and wounds. Some eat the cooked leaves and rhizomes, but there are irritating calcium oxalate crystals present in fresh material.

Begonia plebeja

Begonias
Begonia species

Family: Begoniaceae

Not long ago, a Costa Rican neighbor who loves to garden took me through her backdoor into her laundry room. The *pila* (a utility sink) is no longer useable for washing since it is full of begonias. Every shelf and window sill is crammed with plants with leaves that are small or large, bizarre and crinkly, and/or beautifully blotched with splashes of color. New plants are forming in an assortment of containers, some from leaves simply stuck in the soil. Her burgeoning begonia collection appears to be taking over!

Begonia enthusiasts everywhere are passionate about their plants, and they exchange information and plants through organizations such as the American Begonia Society. Members of the ABS grow *Begonia* hybrids with imaginative names such as 'Careless Whisper', 'Avalanche', 'Christmas Candy', and 'Curly Face'. There are more than 10,000 hybrids and garden cultivars. Rex begonias are especially popular for their colorful and fancy foliage. One can propagate begonias by stem or leaf cuttings; even sections of leaves can produce new plants.

In their native habitats, begonias grow on the forest floor, cling to rocky slopes, or cascade down from tree branches, although they are perhaps more often seen decorating patios than growing in the forest. Some are small and compact, while others are several meters tall. Begonias grow at sea level and at elevations as high as 10,000 ft (3,000 m).

The begonia's stems are jointed and often fleshy. The alternate, frequently asymmetrical leaves, which exhibit a variety of colors and textures, may be lobed or toothed. The capsular fruit is three-lobed, with a wing or spur, and has many, minute seeds.

The flower color is usually pink or white, although some species bear red, orange, or yellow flowers. The blossoms, which are often small and delicate, are either male or female. Pollination occurs through deceit, since the yellow stigmas of female flowers look very similar to the anthers of male flowers. This fools pollen-seeking bees into visiting female flowers, and, in doing so, they often pollinate the flowers with pollen they have picked up from male flowers.

Begonia is the largest genus in the Begoniaceae family, which comprises about 1,400 species in tropical America, Africa, and Asia. Another genus, *Hillebrandia*, occurs only on the Hawaiian islands. In Costa Rica there are around thirty-four native *Begonia* species. Some, such as *Begonia multinervia*, a beautiful plant whose leaves have a wine red underside, have potential as garden or potted plants. Certain begonias have edible flowers, stems, leaves, or rhizomes. In the Mexican state of Puebla, country people eat six different species.

Begonia hirsuta

Begonia estrellensis

Begonia heydei

Bird of paradise, Ave del paraíso
Strelitzia reginae

Family: Strelitziaceae

The well-known bird of paradise, a cousin of the heliconias, is a horticultural favorite. It grows in clumps 3 to 4 ft (1 to 1.5 m) tall and has gray-green leaves and boat-shaped green bracts with reddish borders. The showy orange parts of the flower are sepals; the blue parts are petals that form a keel, enveloping the long anthers and style. A shorter third petal covers the nectary. In its natural habitat, bird of paradise is pollinated by sunbirds that land on the sturdy blue petals. These form a perch that allows a bird to reach the nectar at the base of the flower. The weight of the bird causes the petal lobes to spread apart; this exposes the pollen-bearing anthers, which contact the bird's feet. The Strelitziaceae family contains some very large plants, some taller than 35 ft (10 m), including white bird of paradise (*Strelitzia nicolai*), traveler's tree (*Ravenala madagascariensis*), and a South American species, *Phenakospermum guyannense*. The family name honors Charlotte von Mecklenburg-Strelitz, the queen to England's George III.

Bird's-nest anthurium, Tabacón
Anthurium salvinii

Family: Araceae

This epiphyte—which can also grow on steep, rocky slopes—has thick, crowded roots. Found in Mexico, parts of Central America, and Colombia, it is one of the largest anthuriums in Costa Rica. It often appears at about 3,300 ft (1,000 m) in moist to wet forest on both slopes. It is common in the Tilarán mountain range, where it is frequently used as an ornamental plant. Overlapping leaves radiate out from the center of the plant; the leaf blades are 3 to 6 ft (1 to 2 m) long. An 18 in (45 cm) purple or green-tinged purple spathe subtends a pendant, lavender spadix on a long stalk. When it is in fruit, the spadix lengthens and is covered by many red berries. The Spanish name, *tabacón*, is given to a variety of plant species whose leaves resemble those of tobacco. The rosette of leaves collects precipitation and leaf litter and directs it to the center of the plant.

Bougainvillea, Veranera
Bougainvillea species and cultivars

Family: Nyctaginaceae

The brilliance of crimson and magenta bougainvilleas on a sunny, dry-season day is such that the plants seem to shimmer. While these vibrant colors are popular, subtler shades of orange and pink are not uncommon. Bougainvillea is a shrub or liana with alternate smooth-edged leaves whose shape and degree of fuzziness varies with the species or cultivar. Each of the colorful papery bracts, which are arranged in threes, has a narrow, five-lobed, tubular flower, about 1 in (2 cm) long, attached on its inner surface. Hummingbirds and butterflies visit the flowers, and the bracts probably play a role in the wind dispersal of seeds in naturally growing plants. Various species have their origin in South America, mainly in Brazil—the common and scientific names come from Louis de Bougainville, the eighteenth-century French navigator who discovered the plant in Rio de Janeiro—and many cultivars and hybrids have been developed and exist in gardens throughout the world. In Costa Rica, bougainvilleas are very common in yards and on hotel grounds throughout the country, from low to fairly high elevations. The plant is also called paper flower.

Brazilian red cloak, Pavoncillo rojo
Megaskepasma erythrochlamys

Family: Acanthaceae

The family Acanthaceae includes many ornamentals, but it is also one of the most common families in the understory of Costa Rican tropical forests. Although the Brazilian red cloak originated in Venezuela, it is now in cultivation in many tropical countries; in Costa Rica, it appears in yard plantings at mid elevations. A very showy hedge plant that flowers prolifically, it is easy to grow since cuttings root readily. This shrub reaches a height of 6 to 13 ft (2 to 4 m). Its 10-inch-long (25 cm) spikes have magenta to red bracts with tubular, white two-lipped flowers. Like in many other Acanthaceae species, when the club-shaped capsule splits open, the four seeds are shot out explosively. This species is also known as red justicia.

Typical tank bromeliad.

Guzmania nicaraguensis

Bromeliads, Bromelias

Family: Bromeliaceae

This tropical plant family is one of the easiest to recognize. With their rosettes of straplike leaves, in which they collect water and organic debris, bromeliads look like pineapple tops perched on tree trunks and branches. Not all bromeliads are epiphytes, however; some inhabit rocky ridges in dry forest, while others flourish on the ground in wet or dry forest. A few, such as Spanish moss (*Tillandsia usneoides*), have atypical growth forms. Species vary in size from a few inches to 35 ft (10 m). Estimates of the total number of species in the family vary, up to about 2,400, with many ornamental cultivars. With one exception (*Pitcairnia feliciana* of West Africa), this is a neotropical family. About 200 species occur in Costa Rica.

The flowering stalk of a bromeliad, which forms in the center of the leaf cluster, bears flowers that may be purple-black, white, yellow, orange, violet, or pink. Hummingbirds visit many species, while bats or insects pollinate others. Bracts and regular leaves often add bright red to the display. Specialized hairs on the leaf surface, which have the form of stalked scales, play an active role in water and nutrient absorption. The leaves may be narrow and covered with grayish scales, or they may be wider, less scaly, and channeled. Fruits are either berries eaten by birds and mammals or capsules with tufted or winged seeds that are wind-dispersed. The pineapple, with its multiple fruit, is an oddity in the Bromeliaceae family. Bromeliads typically develop side shoots—"pups" in gardeners' language—that take over after the main plant has flowered and died.

Bromeliads are popular with tropical gardeners because they combine continuously handsome foliage and interesting, often brightly colored, inflorescences. The plants also have a predictable compact form and are fairly easy to grow. Many do quite well as houseplants, even when neglected.

Besides pineapple, which is a source not only of food but also of proteolytic enzymes, humans use a few other bromeliad species. Spanish moss is used as a stuffing for furniture, and some *Aechmea* and *Ananas* species provide textile fiber. In Costa Rica, people use the fruits of the wild pineapple or *piñuela* (*Bromelia pinguin*) in jams, desserts, and drinks.

Bromeliads that accumulate water and leaf detritus in the tanks formed by their tightly overlapping leaves provide habitat and/or a breeding ground for protozoa, aquatic insects, and frogs. Some species that live in drier climates have sheathing leaf bases that bulge out, forming hollows where ants can nest. The feces, food waste, or organic garbage of these insects and other inhabitants probably contribute to meeting the plants' nitrogen needs.

Aechmea mariae-reginae

Tillandsia usneoides

Butterfly weed, Viborana
Asclepias curassavica

Family: Apocynaceae

The butterfly weed is a southern relative of the North American milkweeds that many people associate with the monarch butterfly. Adult butterflies sip nectar from flowers, while the striped monarch larvae feast on the leaves. A variety of butterfly species, as well as bees, wasps, and beetles, visit the butterfly weed for nectar; in addition to monarch caterpillars, the plants also host aphids and milkweed bugs. Many *Asclepias* species contain poisonous cardenolides that make the caterpillars, and the butterflies they become, unpalatable to predators. The butterfly weed, or blood flower, is found in Mexico and throughout Central and South America,

as well as in Florida and the West Indies. It has been introduced to the Old World tropics as an ornamental. In Costa Rica, it is found throughout the country, from sea level to 6,600 ft (2,000 m) on both slopes, in second growth forests and pastures, as well as along streams and roadsides. The plant is usually less than 3 ft (1 m) tall, with opposite leaves and white latex. The flower structure is complex, with a "skirt" made up of five red-orange petals, above which is an orange-yellow corona made up of cups and horns, the cups holding nectar. The pods split open to release many small flat seeds with silky tufts that float in the wind.

Canna lily, Platanilla
Canna x *generalis*

Family: Cannaceae

The overall size, as well as flower and leaf color, of the canna lily vary because of the number of cultivars that have been developed. The plants have broad leaves that clasp the stem. The flowers are often red, orange, or yellow spotted with red, but may be other colors. The fruit is a capsule that wears thin over time to expose and release hard, round seeds. In Costa Rica, the canna lily is found in gardens at various elevations. The less flamboyant Indian shot (*Canna indica*), a native neotropical species that has smaller red-orange flowers, is also seen in cultivation. This plant's hard, round seeds are used as beads but also have potential as a substitute for lead shot. *C. indica* may actually be one of the parent species of the fancier *Canna* x *generalis*, which is the result of hybridization.

Chenille plant, Rabo de gato
Acalypha hispida

Family: Euphorbiaceae

Many of the 450 or so species of *Acalypha* that occur in tropical areas around the world grow in disturbed habitats. Around twenty species are native to Costa Rica. The chenille plant—sometimes called red-hot cattail—is native to New Guinea and the Malay archipelago. It is a tall shrub with oval-shaped, toothed leaves that are somewhat hairy or bristly. Except for their color, the female inflorescences, which can be more than 12 in (30 cm) long, truly look like furry cat tails. The branching styles of the many small flowers, rather than the petals, are what give the red color to the pendant "tails." Male and female flowers are on separate plants; since the male flower spikes are tiny and inconspicuous, only the female plants are cultivated in gardens. The fruit, a three-parted capsule, is not usually seen in garden specimens. These shrubs grow well in partial to full sun. They may be grown in containers or as a hedge; new plants can be started by cuttings. In Costa Rica, this garden ornamental grows from sea level to mid elevations. Care should be taken when handling the plant, since sap from cut plant parts may irritate the skin.

Cochineal, Tuna
Nopalea cochenillifera

Family: Cactaceae

This terrestrial cactus, which can grow to more than 13 ft (4 m), sometimes has a trunk of 8 in (20 cm) or more in diameter. Its large, flat green pads are usually spineless, but if spines are present they are few and about a half-inch (1 cm) long. The flowers are borne singly, often along the edge of the pad. The lower part of the flower is broad, firm, and green, while the upper part is rose-colored with peach tones; many yellow and pink stamens extend beyond the petals. The plant, also called *nopal* and *caite,* has a many-seeded red fruit that, along with the young pads, is edible. The cochineal's origin is unknown, but it is probably from southern Mexico. It is cultivated in many areas, historically for red dye but now as an ornamental. In Costa Rica, it appears from lowlands to about 4,600 ft (1,400 m). This cactus is the host plant of the well-known cochineal insect, the source of red dye dating back to pre-conquest Mexico. The dye, which was first sent to Spain in the early 1500s, is prepared by cooking the insects, then drying them. The plant has also been used as animal forage and as a hedge plant.

Coffee, Café
Coffea arabica

Family: Rubiaceae

Grano de oro (grain of gold), as it is known in Costa Rica, arrived in 1808 and is now one of the top export products of the country. People in Ethiopia, the plant's native land, supposedly chewed on the leaves and berries; but it was on the Arabian Peninsula, on the other side of the Red Sea, where cultivation began and the beverage was developed, possibly as long as a thousand years ago. The plant can grow to 26 ft (8 m), but is usually kept under 6 ft (2 m). Its opposite leaves are dark green and shiny. The fragrant, five-lobed white flowers come directly off the stem and often appear in large bunches. The fleshy, inch-long (2 cm) red fruit has two seeds. Coffee flowers from March through May, and the fruits ripen and are picked between November and February. After the harvest, a machine removes the fruit pulp, and the seeds are washed and left to ferment for up to a day, thus enhancing the flavor and aroma. The seeds are then sun-dried, the remaining skin is removed, and the coffee "beans" are roasted. Coffee's stimulating effect results from the ability of caffeine to block adenosine, which occurs naturally in our bodies and has a calming effect. When it is blocked, we become—and stay—wired and feel more energetic and less depressed. Caffeine overdose can lead to the jitters, gastrointestinal irritation, rapid or irregular heart-beat, and anxiety attacks.

Corn plant, Caña india
Dracaena fragrans

Family: Dracaenaceae

Native to Africa, the corn plant, or dracaena, is planted in a range of climates in Costa Rica, where it is common as living fences along stretches of the Inter-American Highway, often in coffee plantations. The plant should look familiar, since it appears in office reception areas around the world. Its ease of care makes it an ideal interior foliage plant; a NASA study lists it as one of the top ten plants most capable of removing toxins from indoor air. In Costa Rica, cuttings readily grow into living fences that help stabilize soil on steep slopes. The corn plant is sometimes harvested as a horticultural export crop, since it can easily be shipped as stem sections that are propagated later in greenhouses. The corn plant has a leafy stem, usually not branching, that can grow to 20 ft (6 m). It has arching straplike leaves that clasp the stem; the cultivar 'Massangeana' has yellow striped leaves. The plant has dense groups of short-tubed, fragrant, white-to-violet flowers, followed by red-orange fruits.

Costa Rica nightblooming cactus, Pitahaya
Hylocereus costaricensis

Family: Cactaceae

Although plants in the Cactaceae family are most often associated with arid climates, most of the forty species of cactus in Costa Rica grow in moist and wet forests, and the majority of them are epiphytic. Many bloom nocturnally and have scented flowers, principally white, that are pollinated by hawkmoths and bats. Other species, with flowers of various colors, bloom during the day and attract hummingbirds and insects. The epiphytic cactus with the most impressive flowers is the Costa Rica nightblooming cactus (*Hylocereus costaricensis*), a species with branching, 3-winged or triangular stems. If you travel through the Guanacaste region of Costa Rica, you will see this cactus hanging from branches of older trees. It occurs from Nicaragua to Panama, from lowland areas to around 4,420 ft (1,350 m). The fragrant flowers, which appear in the wet season, open to a diameter of about 12 in (30 cm). The bright magenta fruits are edible; *pitahaya*-flavored ice cream is sometimes sold at the Pops ice cream chain in Costa Rica.

Costa Rican bamboo palm, Pacaya
Chamaedorea costaricana

Family: Arecaceae

Chamaedorea, with about ninety-five species, is the largest palm genus in the neotropics; thirty-one of these species are found in Costa Rica. *C. costaricana* is the only species native to Costa Rica that forms conspicuous colonies. The trunks, or stems, of the Costa Rican bamboo palm are slender and green. In the forest understory, where they tend to grow, the lanky stems, which have prominent nodes, reach toward the canopy and may look like bamboo to the casual observer. The smooth, pinnately compound leaves have segments that are narrow and slightly S-shaped. Yellowish flowers appear in branching clusters, and the fruiting stem is orange with dark olive fruits that are black when ripe. The plant is found from Mexico to Panama. In Costa Rica, it grows on both slopes, from about 2,000 ft (600 m) to more than 6,600 ft (2,000 m). It is more widespread on the Pacific slope. This plant is attractive as an ornamental, and it is especially popular in the Costa Rican town of Pacayas, east of San José. Bees, beetles, and flies visit the bamboo palm's flowers, while other *Chamaedorea* species are wind-pollinated. The Spanish name *pacaya* is given to many species of *Chamaedorea*.

Crepe ginger,
Caña agria
Cheilocostus speciosus

Family: Costaceae

While not a true ginger, the crepe ginger is a member of the Costaceae, a family closely related to, and sometimes lumped with, the ginger family, Zingiberaceae. Crepe ginger and closely related *Costus* species are also known as spiral gingers because of their corkscrew stems. The plant grows to 10 ft (3 m) and has sheathing leaves that spiral around the stem. These leaves have a tapering tip and a flannel-like underside. Carpenter bees pollinate the white flowers and ants are often seen visiting nectar-producing glands on the red bracts of the conelike inflorescence. Originally from Southeast Asia, India, and New Guinea, in Costa Rica this plant usually appears in warm, wet or moist lowlands, sometimes escaping from cultivation.

Croton, Crotón
Codiaeum variegatum

Family: Euphorbiaceae

Full or morning sun produces the best color in these plants, which are useful for providing constant color in landscaping. They are used as hedges and are widespread as ornamentals, sometimes as potted plants, in the lowlands of Costa Rica, but may also be seen up to elevations of 4,900 ft (1,500 m). Originally from South India, Sri Lanka, and islands of the Pacific, they are now cultivated throughout the tropics. The croton, also called *palo de oro* in Spanish, is usually maintained as a shrub, but it may grow to more than 10 ft (3 m) tall. Any plant you see that looks like a variation of the *Codi-*

aeum variegatum is probably some cultivar of this species. The alternate, glossy leaves are quite variable in size, shape, and color. They may be lobed, have a ruffled, or even stringy, appearance, and may be twisted. Blotches or streaks of white, pink, orange, yellow, bronze, and red deck the leaves—some of the colors change as the plant ages. Small, white flowers appear in long spikes. The Euphorbiaceae family includes poinsettia, cassava, and castor bean.

Dumb cane, Lotería
Dieffenbachia species

Family: Araceae

The name dumb cane comes from the fact that ingesting this plant impairs speech, not to mention breathing and swallowing; calcium oxalate crystals and other unknown substances (possibly toxic proteins) in the leaves and stem produce burning and swelling of the mouth and throat. Although death in adults is unlikely, resultant fatal choking has been reported. Dumb cane species comprise a well-known group of houseplants that are frequently seen in shopping malls and offices; there are many variegated cultivars. Although the exact number of species in Costa Rica is not clear, there are at least a dozen. These fleshy plants are often about 3 ft (1 m) tall, but sometimes grow to more than 6 ft (2 m). They have a milky sap and, in some species, the crushed plant gives off a skunklike odor. The leaves are sometimes mottled or have a distinct white or yellow stripe. The lower part of the green spathe encloses the female flowers, but the upper part is open and hoodlike, exposing the male flowers. The ripe, orange to red fruits make a showy display when the spathe splits open. In Spanish, *Dieffenbachia* species are known as *lotería*, *sainillo*, and *rayo de luna*.

Dwarf poinciana, Malinche
Caesalpinia pulcherrima

Family: Fabaceae
Subfamily: Caesalpinioideae

The origin of the dwarf poinciana is unclear; it appears to be native to Mexico, the West Indies, and parts of Central America, however some sources say it is native to Asia. It is often cultivated as an ornamental—it makes an attractive flowering hedge—and in Costa Rica it is planted from low to mid elevations, escaping at times and growing in disturbed habitats. A shrub or treelet, this plant reaches a height of 16 ft (5 m); its trunk and branches may have prickles. Large compound leaves are complemented by flowers with five crinkled petals. The blossoms are first red-orange with a yellow border, then become entirely red, although the color is variable. One variety, for instance, is all yellow. The colorful flowers attract nectar-seeking butterflies, which are the chief pollinators, and hummingbirds. The plant contains hydrocyanic acid and is potentially toxic, but it is also medicinal. Research indicates some antifungal and antibacterial activity in the plant. A variety of folk uses range from cleaning teeth to treating venereal disease. Dried, crumbled flowers supposedly kill insects.

Firebush, Zorrillo real
Hamelia patens

Family: Rubiaceae

The firebush, also known as *zorrillo real* and *coralillo*, occurs from sea level to 5,250 ft (1,600 m) and is common along roadsides in some parts of the Caribbean region of Costa Rica. It ranges from South Florida, the West Indies, and Mexico, to Central America and parts of South America. This shrub or small tree, which may reach a height of 16 ft (5 m) or more, is used as a medicinal plant in various parts of its range. The plant has antibacterial and antifungal properties and is rich in alkaloids. As a bath or poultice, it is applied to various skin ailments; a leaf tea is taken for worms and to relieve fever. Often found in sunny forest gaps or other disturbed habitats, firebush is a great candidate for native-species landscaping. Its flowers attract hummingbirds and butterflies, and various birds eat the fruits. Visiting hummingbirds often transport tiny mites that hop off into the flower, where they ingest some pollen and a good deal of nectar. Red-orange tubular flowers, with five corolla lobes, appear at the ends of the twigs. The fruits, which contain tiny seeds, occur in clusters that consist of dark red immature and purple-black mature fruits.

Flame of the woods, Flor de fuego
Ixora coccinea

Family: Rubiaceae

This popular hedge plant has a variety of common names—flame flower, burning love, jungle flame, and *cruz de Malta*. Usually maintained as a shrub under 6 ft (2 m), the flame of the woods has dense foliage with opposite, stalkless leaves that attach directly to the stem. The flowers, which appear continuously, are tightly packed in rounded heads and are pinkish or red, with four-lobed corollas. In its native habitat, the narrow-tubed flowers are probably pollinated by butterflies. The origin of the plant is India, but it is found in cultivation throughout Central and South America, the West Indies, and Madagascar. In Costa Rica, it appears in sunny gardens and parks, from the lowlands to mid elevations. *Ixora casei* is another cultivated species sometimes confused with *I. coccinea*; it is distinguished by leaves that have short stalks (as opposed to none) and seven to twelve pairs of secondary veins (as opposed to five or six pairs).

Flame vine, Triquitraque
Pyrostegia venusta

Family: Bignoniaceae

Originally from Brazil, Paraguay, Bolivia, and northeastern Argentina, the flame vine, or firecracker vine, is found in cultivation throughout the neotropics and is sometimes naturalized. New plants are easily grown from stem cuttings; the flame vine is a popular ornamental at middle elevations, especially in the Central Valley. Its typical habit is cascading down walls, scrambling along fences, and covering rooftops; it will also climb trees. This liana has opposite, compound leaves, and climbs by way of tendrils that grow from the leaves. The flowers, which grow in large, usually pendulous, clusters, have five lobes that curl back at the ends; these tubular, red-orange flowers attract hummingbirds. The fruit is a long, thin capsule.

Glorybush, Nazareno
Tibouchina urvilleana

Family: Melastomataceae

In tropical and semi-tropical regions of the world, glorybush, a large shrub that can grow to around 16 ft (5 m) tall, is prized for its foliage and its 4-in-diameter (10 cm) intense violet flowers. Most parts of the plant, including the leaves and square stems, are covered with hairs or bristles. The pubescence on the leaves feels and looks like velvet, and the venation has a netted appearance. Two pinkish bracts enclose the flower bud; the five petals as well as the ten stamens are purple. The glorybush—also called princess flower, *nazareno*, and *príncipe negro*—is native to southern Brazil. It is grown as an ornamental in various parts of the world, including the southern United States. In Costa Rica, it is planted from around 4,300 to 7,500 ft (1,300 to 2,300 m). The Melastomataceae family is a common component of neotropical rainforests, where several thousand of its members occur. The family is easy to identify by the netted leaf-vein pattern.

Golden dewdrop, Duranta
Duranta erecta

Family: Verbenaceae

The golden dewdrop originated in tropical America; it is widespread as a cultivated and naturalized plant from southern Florida and Texas to Argentina. It has been introduced to Africa, Asia, Hawaii, and Australia. In Costa Rica, where it is not native, it appears in parks, hedges, and yards, especially at mid to higher elevations. This ornamental—also called *once de abril*—attracts a wide variety of bees, moths, and butterflies, as well as hummingbirds. One scientist has observed more than ninety species of bees and lepidoptera visiting the flowers, which have a sweet, candy-like scent in the late afternoon and evening. The flowers have five rounded, lavender petals and a white center. The plant grows as a shrub or small tree to 20 ft (6 m) tall. It has bunched leaves that have a wedge-shaped base. The yellow-orange fruits, which hang in stringlike clusters, are toxic; ingesting them results in fever, nausea, vomiting, convulsions—and sometimes death. Despite this, the plant is used medicinally in Guatemala and Mexico to reduce fever and as a stimulant. The juice of the fruit kills mosquito larvae.

Golden shrimp plant, Olotillo
Pachystachys lutea

Family: Acanthaceae

The golden shrimp plant can be grown from cuttings and is cultivated in tropical gardens as well as in containers. Originally from Peru, this species is planted in many gardens in the Central Valley of Costa Rica. It has elliptical leaves and erect, terminal inflorescences composed of overlapping golden-yellow bracts in four rows. The two-lipped flowers are white, and the fruit is a capsule that splits open when mature. The true shrimp plant, *Justicia brandegeana*, which comes from Mexico, is aptly named because the overlapping pink bracts of its curved inflorescence look like segments of a shrimp's abdomen.

Golden trumpet, Bejuco de San José
Allamanda cathartica

Family: Apocynaceae

A liana or shrub with milky latex, the golden trumpet, or yellow allamanda, has glossy leaves that are either opposite or in whorls of three or four. Its bell-shaped yellow flowers have five rounded (to lobed) overlapping petals. The fruit is a round, green, spiny capsule; the winglike edges of the seeds suggest they are dispersed by the wind. Considered to be from northeastern South America, the golden trumpet is planted as an ornamental in Costa Rica, from sea level to 3,300 ft (1,000 m) or higher. It sometimes grows in abandoned pastures or along Caribbean waterways, where it appears that it might be native. Its family includes rosy periwinkle and frangipani. The plant is poisonous, and the species name *cathartica* implies its effect—stimulating evacuation of the bowels. The leaves, latex, and bark may cause a rash.

Heliconia bihai

Heliconia psittacorum

Heliconias, Platanillas
Heliconia species

Family: Heliconiaceae

The colorful bracts of the heliconias, like the dazzling flowers of the bird of paradise, make these plants choice ornamentals for tropical gardens and flower arrangements. In gardens, heliconias are easily grown from small plants or rhizomes. Although their general form and leaf shape are similar to that of the bird of paradise and the banana, members of the Heliconiaceae family have a distinctive flower display. Many bracts of brilliant colors enclose the much smaller green, yellow, or orange flowers, while in the bird of paradise it is the flowers, not the single bracts, that attract attention. The flashy colors function as flags that attract pollinating birds—hummingbirds in the case of heliconias, and sunbirds for bird of paradise.

There are more than thirty-five *Heliconia* species native to Costa Rica and between 200 and 250 species in the world, with perhaps as many, if not more, forms or cultivars. Species of *Heliconia* from the New World now appear in cultivation in Asia, Africa, and Hawaii. Commercial growing for cut flowers, which began in Hawaii and in greenhouses in Holland and Germany, has spread to other regions, including Costa Rica, where growers mainly export heliconias to the United States market.

A close look at a heliconia flower reveals three sepals and three petals that are held together to form a tube. The variation in the length and shape of flowers of different species demonstrates how these plants have coevolved with hummingbirds. The curved flowers are adapted for pollination by hummingbirds such as the hermits, which have long, curved bills; the shorter straight flowers are suitable for visits from a variety of non-hermits. Different flower species dab pollen onto distinct parts of a bird's bill or head, making the transfer of pollen to the stigma of the next individual of that species likely to occur.

Like all tropical plants, heliconias are part of a web of interactions within their environment. Many have erect boat-shaped bracts containing liquid that houses protozoa and insect larvae, especially those of flies. Mites that commonly inhabit the flowers hitch rides to other plants on hummingbird bills. The fruit, which turns blue when ripe, attracts manakins, flycatchers, motmots, and other birds. Indigenous and country people use heliconia leaves as food wrappers and for thatching.

Heliconias may be seen in most natural areas in Costa Rica, as well as in botanical gardens such as the Wilson Botanical Garden in San Vito, which has a superb collection. The greatest number of heliconia species occur at mid elevations. In natural settings, sun-loving species grow along streams, or in gaps in the forest where there has been a tree fall or landslide. Some of these are almost weedy, growing along roadsides and in second-growth situations.

Heliconia wagneriana

Heliconia pogonantha

Hibiscus, Clavelón
Hibiscus rosa-sinensis

Family: Malvaceae

Beautiful and easy to propagate—cuttings stuck in the ground take readily—hibiscus is one of the most common hedgerow plants in Costa Rica. There are more than 200 species of *Hibiscus*, and thousands of cultivars (including red, white, and peach-colored cultivars of *H. rosa-sinensis*). The genus *Hibiscus* is found in tropical, sub-tropical, and some temperate regions of the world; some species are native to Hawaii. The exact origin of *Hibiscus rosa-sinensis* is unclear, but it is somewhere in tropical Asia. *Amapola*, as the hibiscus is also called in Spanish, is a shrub or tree with alternate, shiny, toothed leaves, and a five-petaled flower. A column protruding from the center of the flower is topped by a starlike arrangement of five styles, with many stamens below. Crushed hibiscus flowers yield a juice that has been used as shoe polish, mascara, and hair dye; flower and leaf decoctions are used in various places to treat flu, coughs, and asthma. The flower calyces of rosella (*H. sabdariffa*), which is cultivated in Costa Rica, is a component of Red Zinger brand tea. Other species in this family include okra (*Abelmoschus esculentus*) and marshmallow (*Althaea officinalis*), the source of genuine marshmallow!

Impatiens, Chinas
Impatiens walleriana

Family: Balsaminaceae

Many people are surprised to discover that the prolific impatiens, also called busy Lizzie, is not native to Costa Rica, since it is certainly at home in many parts of the country. Originally from the mountains of East Africa, it is common in Costa Rica from lowlands to 5,250 ft (1,600 m); it is planted in gardens, but it has escaped to open areas such as road- and streamsides. A native species, *Impatiens turrialbana*, which has red-orange flowers, is found at 4,300 to 8,200 ft (1,300 to 2,500 m) in Tapantí, on slopes of Turrialba and Irazú volcanoes, and in parts of the Talamanca mountains. Another member of the Balsaminaceae family is the North American jewel weed (*I. capensis*), also called touch-me-not. *Impatiens walleriana* is an herbaceous plant with a juicy stem that grows to 3 ft (1 m). Its bilaterally symmetrical flowers, which appear in a variety of colors (red, pink, orange), have an arching, narrow spur, five petals, and purple pollen. The flowers attract both butterflies and hummingbirds. The ribbed, bulging, green capsules burst open upon the slightest disturbance and shoot out small seeds. Once you have seen a mature, turgid capsule burst and disperse its seeds, the urge to squeeze them is irresistible.

Kohleria
Kohleria spicata

Family: Gesneriaceae

This is one of the wild roadside plants of the African violet and gloxinia family (Gesneriaceae). The kohleria's rhizomes allow it to go into a dormant phase during particularly dry spells, sprouting anew when moisture is available. As with many native gesneriads, the kohleria has potential as an ornamental and is a good candidate for use in native plant gardens in Costa Rica. It is also great for attracting hummingbirds. Ranging from Mexico to Ecuador, it is widespread in Costa Rica, from the lowlands to elevations above 5,250 ft (1,600 m). It is seen on steep road and stream embankments, and occurs from Guanacaste south to the Osa Peninsula. The whole plant, including the flowers, is covered with short, soft hairs (red on the stem). The plant has tooth-edged opposite leaves that sometimes appear in whorls of three. Kohleria's tubular flowers are brilliant red-orange, yellowish in the throat and with streaks or rows of red dots on the five corolla lobes.

Lantana, Cinco negritos
Lantana camara

Family: Verbenaceae

Considered an ornamental by some, lantana is a scourge in parts of the world, such as the Old World tropics, where it has escaped and become a pest. Lantana ranges from the West Indies and the southern United States to northern South America, and is planted as an ornamental and naturalized in South America, the Pacific Islands, Australia, New Zealand, and parts of Asia. In Costa Rica, lantana occurs naturally in old pastures and second-growth, or is cultivated in yards, up to 6,600 ft (2,000 m). Also called red sage, yellow sage, *cinco negritos*, and *soterré*, lantana is a shrub with hairy square stems that may or may not be prickly. The toothed, opposite leaves are moderately scratchy above, with soft hairs on the underside. All plant parts give off a pungent odor. The flower heads have orange-red (or pinkish) buds in the center; the fresh flowers are yellow, and old flowers turn orange to red (or pink). Butterflies and hummingbirds visit the flowers for nectar. The fruits ripen to a shiny, metallic blue-to-black. Unripe fruits are known to cause vomiting, weakness, and respiratory and circulatory problems that may result in death. This is one of a number of plants that has a range of internal and external folk-medicine uses but, at the same time, contains known toxins and should be handled with care—literally, since it may cause dermatitis. Some domesticated animals are adversely affected by eating the plant.

Lobelia, Caragallo
Lobelia laxiflora

Family: Campanulaceae

While many lobelias have ornamental qualities, *Lobelia laxiflora*, perhaps because of its scraggly appearance, does not appear in gardens. It nevertheless merits a place in wildlife gardens since the flowers attract hummingbirds. A species of hummingbird flower mite, *Tropicoseius chiriquensis*, lives in the flower tubes and hitches rides on birds' bills to fresh flowers. This lobelia appears from southern Arizona to Colombia, and in Costa Rica it is seen in disturbed areas such as roadsides, embankments, and old pastures, from about 3,300 to 8,500 ft (1,000 to 2,600 m). This branching, latex-producing plant, which has alternate leaves, is 3 to 6 ft (1 to 2 m) tall, often clumped. Its many flowers attach at the leaf or bract axils toward the top of the stems. The yellow-tinged, red-orange flowers are basically tubular, with two main lobes, one being slit above the anther tube and the other forming a broader lower lip.

Marriage vine, Volcán
Solanum wendlandii

Family: Solanaceae

Once thought to be endemic to Costa Rica, where it is widely cultivated and found at altitudes of 2,300 to 5,600 ft (700 to 1,700 m), the marriage vine is now known to grow in other parts of Central America. Sometimes called potato vine, this plant is related to the potato, tomato, eggplant, and bell pepper. It is a spiny climber that ascends high into the canopy. The underside of its leaves, as well as the stem, are armed with hooked prickles. Its branched inflorescences comprise many violet-blue flowers that are buzz-pollinated by a variety of bees, including bumblebees (*Bombus*). To collect pollen, the bees must vibrate the anthers, which have terminal pores where the pollen billows out.

Ipomoea trifida

Morning glories, Churristates
Ipomoea species

Family: Convolvulaceae

The majority of the 500 species of morning glories occur in the tropics, with close to fifty species in Costa Rica. Most *Ipomoea* species are twining vines that climb without tendrils. Usually the plants contain milky latex. The simple, alternate leaves may be lobed. The typically funnel-shaped flowers often have stamens of different lengths, and the capsular fruit has four sections. With such similarity within the genus, it is the sepals, at the base of the flower tube, that distinguish *Ipomoea* species. In *I. trifida* for ex-

ample, the outer sepals are shorter than the inner sepals, and they are hairy, with a narrowed tip.

Most morning glories open at dawn, are pollinated by bees, moths, and skippers, and wilt by late morning or afternoon. The pantropical moonflower (*I. alba*) is an "evening glory" that opens at dusk and has a wonderful nocturnal perfume that draws in hawkmoths.

From November to February—the peak flowering time for morning glories—early morning travelers in Costa Rica will be treated to an array of blossoms along roadsides. Some trail along the ground, while others scramble over shrubs or barbed-wire fences; still others climb into tree canopies. The beach morning glory, *I. pes-caprae*, which grows along the coasts of both the Old and New World tropics, can be seen just above high-tide line in Costa Rica, on both coasts and on Cocos Island. The widespread *I. trifida*, found from Mexico to northern South America, grows in Costa Rican lowlands to 4,900 ft (1,500 m), on roadsides, stream sides, and pasture edges. It is common in Guanacaste, but also occurs south to the Osa Peninsula.

Morning glories have played a variety of roles in the lives of humans. Their colors and forms have pleased gardeners, and the thickened roots of *I. batatas* (sweet potato, or *camote*) have fed millions. Certain indigenous Mexicans have used the divining qualities of the hallucinogenic seeds of *I. violacea* (also called *I. tricolor*) as a way of determining the cause and cure of diseases. The seeds of this species (including garden cultivars such as 'Pearly Gates', 'Heavenly Blue', and 'Flying Saucers'), and of the related *Turbina corymbosa*, contain potentially harmful LSD-like alkaloids. Various morning glory species have been used medicinally; some have purgative qualities. A decoction of the beach morning glory is ingested for rheumatism and kidney ailments; skin ulcers and swellings are treated with poultices or decoctions. A leaf-extract ointment is effective in treating jellyfish stings.

Ipomoea alba

Typical morning glory flower.

A sobralia, or *flor de un día*, orchid.

Orchids, Orquídeas

Family: Orchidaceae

Orchids are at once the most beautiful and the most bizarre plants, exhibiting a variety in flower size, color, and shape that is astounding. They are found throughout the world in a range of habitats. Since hundreds of species have flowers that are just a half-inch (1 cm) or smaller, many are not very conspicuous. Most orchids grow as epiphytes. Those that do not live in the forest canopy grow on the ground or on rocks. Vanilla is an orchid that grows as a high-climbing vine.

The Orchidaceae is one of the largest families of plants in the world, with more than 20,000 species. An even greater number of ornamental hybrid orchids now exist in gardens, florist shops, and live collections around the world. In the tropics, the diversity of orchids is phenomenal. In Costa Rica alone, there are roughly 1,400 species, and some parts of the country are particularly rich; there are about 500 species in the Monteverde region, for example.

One might logically wonder how so many species that look so different can be in the same family. What holds them all together is a distinctive and unique set of flower characteristics. A special petal called the labellum, or lip, is showier than the rest and may be adorned with spots of color, wartlike projections, and frills. The lip is often what attracts the pollinator to the flower, whose pollen is in packets called pollinia. Orchids also typically have seed capsules that split open to release thousands of tiny, dustlike seeds that are wind-dispersed. And some species have pseudobulbs, a special adaptation for water storage.

Pollination systems in orchids are fascinating. Hummingbirds feed on nectar of a few of the species with red, purple, or orange flowers, but other orchids lure insects in with various rewards or use trickery to attract them. Many orchids have no nectar, but offer scent compounds, wax, or oil to their pollinators. Some species mimic other flowers, female insects, rotten meat, or mushrooms. Insects from large hawkmoths to tiny flies are among known pollinators. Male euglossine bees, also called orchid bees, collect the scent compounds of certain orchids in order to make their pheromones.

Because many orchid species are small, rare, and epiphytic—all characteristics that make them challenging to study—much about the life of orchids, their role in the forest, and their interactions with other organisms remains a mystery. As forest destruction and illegal collection and trade continue, some important information will be lost forever.

The species shown here are few in comparison to the size of the family, but they help demonstrate the amazing diversity and beauty of this family.

Ponthieva formosa

A miniature orchid, *Lepanthes*.

A *Masdevallia* species.

Panama hat palm, Chidra
Carludovica rotundifolia

Family: Cyclanthaceae

In Costa Rica, one finds four species of *Carludovica*, or Panama hat palm. The species depicted here, *Carludovica rotundifolia*, grows along the edge of, and within, wet forest from Honduras to Panama. In Costa Rica's Caribbean region, it is found near the coast and in the foothills and mountains; on the Pacific slope, it is found mostly from Carara south to the Osa Peninsula. When in fruit, the plant produces a cylindrical, corncoblike spadix in which the outer layer of the spadix ultimately splits at the top and curls back, revealing a red-orange color on the inside. The yellowish orange fruits are eaten by birds and mammals. This species has a long-stemmed, pleated leaf, divided into four parts, each with fingerlike fringes or segments of about equal lengths. True Panama hats, or *jipijapas* as they are sometimes called, are made from the young leaf fibers of *C. palmata*, whose leaves have longer and more irregular segments or teeth, and a couple of pointed bumps on the blade close to where it joins the stalk. The hats originally came from Ecuador, but they became popular during the building of the Panama Canal, hence the name.

Papaya
Carica papaya

Family: Caricaceae

Papaya fruit is usually eaten raw, but in some Costa Rican households the cooked leaves and/or immature fruits are added to various dishes. A fresh slice with a squeeze of lime juice or a *papaya en leche*, a blender drink with milk, are two of the tastiest ways of eating the fruit. Studies of various parts of the plant show antibacterial, antiyeast, and antifungal properties. Papaya contains lots of vitamin A and C; the redder the flesh, the more vitamin A it has. Papain, a proteolytic enzyme, is another important product of the papaya. While known mostly as a meat tenderizer, papain prevents shrinkage in wool and silk and appears in medicines for digestive ailments. Although native to the New World tropics, it is not clear if the papaya originated in Mexico or in Central or South America. In Costa Rica, it is cultivated throughout the country, not only in large commercial patches, but also as yard plants; it does best in humid areas below 660 ft (200 m). The papaya, or pawpaw, is a soft-wooded tree whose main distinguishing characteristics are an unbranched trunk with conspicuous scars left by fallen leaves. The leaves bunch at the top of the stem, and large, pendulous fruits come right off the trunk. The large palmately lobed leaves have a milky latex, as do the stem and the unripe fruit. Flowers may differ from plant to plant because some individuals are male and some female, but they are generally whitish or yellowish, with five petals. Female flowers have large ovaries and are single or few, while male flowers are tubular and in long clusters. The fruits are melonlike, yellow-orange when ripe, and typically the size of a football.

Passion flower, Granadilla del monte
Passiflora vitifolia

Family: Passifloraceae

According to some religious folklore, components of the fascinating passion flowers depict the story of the crucifixion of Jesus Christ: the three styles are the nails of the cross, the five anthers Christ's wounds, the corona represents the crown of thorns, the tendrils the whips, and so on. Passion flower plants are lianas with tendrils that allow them to climb into the canopy. This particular species flowers near ground level. Its three-lobed leaf has several nectar glands at the base of the leaf stalk. The 5-in-wide (12 cm) flower's five sepals and five petals are intense red, with three green bracts below. The flower's corona is made up of many thin red-to-white filaments. The aromatic fruit is maroon and green, mottled with white. There are more than 500 *Passiflora* species in the world, with nearly fifty in Costa Rica. Their flowers differ in color, size, and shape, and have correspondingly varying pollinators (bees, birds, bats, etc.). *P. vitifolia* appears from Nicaragua to northern South America; in Costa Rica, the plant is widespread, from lowlands to 3,300 ft (1,000 m), and is especially common at La Selva and on the Osa Peninsula. Some passifloras are cultivated for their fruit. For long-wing butterflies (*Heliconius* spp.), the plant serves as food for their caterpillars.

Peace lily, Calita
Spathiphyllum friedrichsthalii

Family: Araceae

The eight species of *Spathiphyllum* that occur in Costa Rica, and the roughly fifty species in the rest of the world, are nearly all similar in appearance. These terrestrial plants have elliptical leaf blades and long-stalked, hooded inflorescences. Several *Spathiphyllum* species and cultivars are widely used as shopping-mall ornamentals. In this particular species, *S. friedrichsthalii*, the spathe, which is white with some green extending into the base on the back, forms an open hood around a fat, cream-to-yellow, cylindrical spadix. The spadix has bisexual flowers and a scent somewhat like a pleasant bathroom disinfectant. The main pollinators are male orchid bees, which visit the flowers for scent compounds that they turn into pheromones. When the plant is in fruit, the spathe turns green; the fruits themselves are green. This species appears from southern Mexico to Colombia. In Costa Rica, it is found in the Caribbean lowlands, from partly open swamp forests to wet roadside and pasture areas. In the southern Caribbean area it is planted in gardens.

Pink velvet banana, Guineo de jardín
Musa velutina

Family: Musaceae

Most banana plants in the world are cultivated for their edible fruit, but a few species have become popular because of the beauty of their fruit or foliage. Originally from India, the pink velvet banana grows outdoors in tropical gardens and indoors in pots; seeds are available online from a variety of sellers. This herbaceous plant grows to about 6 ft (2 m) tall and develops large paddlelike leaf blades. The large flower clusters are upright, with showy red-pink bracts; the flowers are cream-orange to bright orange tinged pink. The small, fuzzy, pink banana fruit is about 2 in (6 cm) long; it splits open when mature, appearing as if someone had peeled the skin back. The seeds are embedded in white pulp. At the La Selva Biological Station in the Caribbean lowlands of Costa Rica and at Wilson Botanical Garden in San Vito, the species has escaped from cultivation and become invasive along rivers and in clearings and young forest. Aggresive exotic species such as this end up costing thousands of dollars in labor and herbicide to ensure that they do not harm native ecosystems. One method of making sure this species' seeds do not spread is to collect them, store them in covered containers for six months, and then bury them deep in the ground.

Poinsettia, Pastora
Euphorbia pulcherrima

Family: Euphorbiaceae

The poinsettia is known in many countries of the world as part of Christmas-season festivities. The plant is native to Mexico and parts of Central America, but is cultivated in many areas. In Costa Rica, it is common in yards from low to medium-high elevations, but a "wild" population was found at about 3,000 ft (900 m) in the central Pacific region in 1995 by collectors from the National Biodiversity Institute (INBio). To northerners, the yard plants they see in Costa Rica appear enormous compared to the horticultural potted-plant counterparts they are accustomed to. The poinsettia is a shrub or small tree that has copious white latex. In addition to alternate, green, long-stemmed leaves that are sometimes lobed, it has red bracts (modified leaves) just below the true flowers. The green, red, and yellow parts in the center of the bracts make up cyathia, small clusters of flowers, that are visited by butterflies, other insects, and hummingbirds. There are many poinsettia cultivars, including some with pink or cream bracts. The poinsettia's toxic latex may cause blistering of skin, and ingesting the plant results in an unpleasant irritation of the mouth and gut, along with vomiting, diarrhea, and, if taken in large amounts, more severe reactions.

Porterweed, Rabo de gato
Stachytarpheta frantzii

Family: Verbenaceae

Stachytarpheta species are great plants for attracting short-billed hummingbirds, butterflies, and moths. Although it is a native, porterweed is easily propagated from cuttings and is more often seen in cultivation. It is a shrub with many branches; soft hairs cover both sides of the opposite leaves as well as the angular branches. The 18-in-long (45 cm) inflorescences have five-lobed, purple flowers with a narrow tube that is light lilac or whitish in the throat. Porterweed occurs throughout Central America; in Costa Rica, the plant grows between 660 and 4,300 ft (200 to 1,300 m), mostly on the Pacific slope, from the central region to the north. Blue snakeweed (*S. jamaicensis*), a less-pubescent plant, is common along the beach and roadsides of the Caribbean and the Osa Peninsula. Some cultivated *Stachytarpheta* in Costa Rica may be *S. mutabilis* (pink snakeweed).

Queen's wreath, Choreque
Petrea volubilis

Family: Verbenaceae

A showy native that often appears as an ornamental in Costa Rica, queen's wreath can be grown as a bush or allowed to spread over an arbor, wisteria-style. The queen's wreath, or purple wreath as it is sometimes called, has opposite, stiff, sandpaper-like leaves. Twelve-inch-long (30 cm) inflorescences, usually hanging, have lilac flowers that may also be bluish or white. The fruit, with one or two seeds, develops enclosed in the calyx, whose five winglike lobes aid in wind dispersal. The author has seen humming-birds visit the flowers, although they are more likely bee-pollinated. Found in the Lesser Antilles, Mexico, and Central America, the queen's wreath has been naturalized in the Old World tropics. In Costa Rica, its native habitat includes forest edges or in overgrown fields up to 3,300 ft (1,000 m) in the Central Valley and on the Pacific slope.

Rattlesnake plant, Bijagua
Calathea crotalifera

Family: Marantaceae

Formerly known as *Calathea insignis*, and also known as *cascabel*, the rattlesnake plant's flowers have a complex trigger system that picks up pollen from probing bees while simultaneously dabbing fresh pollen onto their bodies. The main pollinators are euglossine bees, although other species replace them at higher elevations. The foliage is dark green with a sheen; the leaves are broad, with a light midvein, and a swollen section in the leaf stalk below the blade. Some cooks prefer them over banana leaves for wrapping tamales, while indigenous people of the Talamancas have used them to wrap their dead prior to burial. They also serve well as disposable umbrellas. The flattened golden-yellow (sometimes pinkish) flowering spikes, 16 in (40 cm) long, are made up of overlapping bracts that have an odd plastic texture. The fruit, a yellow capsule, splits open to reveal blue seeds and white arils. The 6- to 13-ft-tall (2 to 4 m) plant grows from Mexico to Ecuador. In Costa Rica, it appears throughout the country in humid areas, from sea level to above 4,900 ft (1,500 m).

Red dracaena, Caña india
Cordyline fruticosa

Family: Asteliaceae

Also known as the good-luck plant or ti plant, this species is cultivated as an ornamental around the world. Red dracaena is native to Southeast Asia and Australasia. In that region, as well as in Polynesia, people eat the root and use the leaves as table cloths, roof thatching, wrappers for food, and, torn into strips, for hula skirts. In Costa Rica, the red draceana appears in yards and gardens in humid environments, from low to mid elevations. The plant is a shrub or small tree with elliptical to lance-shaped leaves that spiral up the stem and cluster at the top. The leaves are all-green or yellow-green suffused with red, maroon, or pink; the pendant, many-branched inflorescence has white or pink flowers followed by red berries. Bees have been seen visiting the flowers, and the size and red color of the fruit suggest that birds disperse the seeds. The plant is also referred to as *Cordyline terminalis*.

Red ginger, Antorcha
Alpinia purpurata

Family: Zingiberaceae

The red bracts of the red ginger add color to Hawaiian leis, and the inflorescences are popular in tropical flower arrangements. In Costa Rica, red ginger is used as an ornamental garden plant in many areas and is especially common in the Caribbean lowlands. It is usually less than 6 ft (2 m) tall, but can reach 13 ft (4 m). The large, sheathing 30-in-long (75 cm) leaves grow singly at nodes along the stem, which is topped by an inflorescence with red bracts and white flowers. The plant is originally from the South Pacific Islands and the Malay Peninsula. Very similar plants with pink bracts or larger flower heads are most likely cultivars of this species. Several other *Alpinia* species are used horticulturally, including shell ginger (*A. zerumbet*).

Rosy periwinkle, Conchita
Catharanthus roseus

Family: Apocynaceae

This common garden ornamental is famous for its alkaloids, vincristine and vinblastine. Researchers working simultaneously in the United States and Canada discovered these compounds to be useful in the treatment of Hodgkin's disease and childhood leukemia, but not in treating diabetes, which is what the plant was used for in Jamaican (and African) folk medicine. At least seventy other alkaloids are present, and ingestion may result in hair loss, muscle deterioration, and damage to the nervous system and the internal organs. Despite its potential harmful side effects, various cultures use it medicinally. The bushy plant has milky latex and opposite, shiny leaves with a light midvein. The five-parted flowers are pink, with deeper pink in the center; the petals are twisted in the bud. Native to Madagascar, the rosy periwinkle is now cultivated from southern Florida and Mexico to South America, as well as in other parts of the world, often escaping into the wild. In Costa Rica, it grows from sea level to 4,900 ft (1,500) m; although it is most often found in yards, it is naturalized in some central and south Pacific coastal areas.

Sensitive plant, Dormilona
Mimosa pudica

Family: Fabaceae
Subfamily: Mimosoideae

Native to the neotropics, but introduced and naturalized in many other parts of the world, including Hawaii and Australia, the sensitive plant appears in Costa Rica from sea level to 8,200 ft (2,500 m) in open areas such as pastures and lawns and along roadsides and rivers. The Spanish common name for this plant, *dormilona*, translates as "sleepy head," a reference to its most striking feature—at night, or when the plant is touched, rained on, or experiences a drastic temperature change, the leaflets fold up and the stalk they are on collapses downward. This reaction may be to evade grazers, but it may also simply reduce transpiration. Its folk use as a sleep inducer may have to do with the doctrine of signatures, a popular belief that plants' physical characteristics signify their potential uses (e.g., liver-shaped leaves cure liver ailments; plants that "go to sleep" bring on sleep). The sensitive plant is a low, spreading herb or shrub with spines that curve backward and alternate, bipinnate leaves. It has conspicuous pink, brushlike flower heads made up of very small calyces and corollas, and long stamens. The fruit is a half-inch- to one-inch-long (1 to 2 cm) segmented pod, with bristles on the edges, that contains several seeds.

Sky vine, Emperatriz eugenia
Thunbergia grandiflora

Family: Acanthaceae

Originally from India, this high-climbing ornamental vine is now seen in tropical areas around the world. In Costa Rica, it appears from lowlands to 4,600 ft (1,400 m); it sometimes escapes from gardens and grows on forest edges, occasionally climbing into the canopy. It is invasive in Hawaii and Australia. Sky vine leaves have sparse, large, jagged teeth. Large, five-lobed, funnel-shaped flowers, lavender-blue with a yellow-white throat, appear in pendant clusters; each blossom lasts just one day. The pollination of sky vine flowers has been studied extensively in southeast Asia. A nectar guide, the purple lines in the throat, attracts large bees; and a washboard path helps them climb into the flower, where pollen is deposited on their backs while they take nectar.

Spineless yucca, Itabo
Yucca guatemalensis

Family: Agavaceae

In Costa Rica, country people propagate *Yucca guatemalensis* by cuttings and harvest the flowers to simmer and fry with herbs and eggs. This is probably an acquired taste since the flowers, a bit like asparagus, are on the bitter side. The spineless yucca is also used in living fences. It appears from Mexico to Guatemala, and in Costa Rica it is cultivated from lowlands to fairly high elevations.

This small tree grows to 26 ft (8 m); its single or branching trunk is thickened at the base. The stiff, pointed leaves are less than 3 ft (1 m) long, and the 25-in (60 cm) inflorescence has branches with many whitish flowers followed by fleshy fruit capsules.

Swiss cheese plant, Costilla de Adán
Monstera deliciosa

Family: Araceae

This well-known ornamental is seen indoors as well as outside. In nature, monsteras begin as vines on the forest floor that head toward a dark place. This is usually the base of a tree, where the plants change their habit and begin to climb toward the sun-filled canopy. The mature dark-green leaves of this thick-stemmed climber are huge, rounded in overall shape, but with deep cuts and holes of various sizes and shapes. The adaptive value of having holes in the leaves may be to allow light to filter down, to mimic insect damage, to shed excess water, and/ or to prevent wind damage. The plant has a bathtub-shaped, whitish spathe and a cream spadix that turns juicy when in fruit. The fruit of *Monstera deliciosa*, which has individual hexagonal berries and takes about a year to mature, is edible—it has a pineapple/banana flavor. Some people appear to be allergic to the fruit. Ranging from Mexico to Panama, the plant is fairly widespread in Costa Rican forests on both slopes, from about 1,600 to 6,600 ft (500 to 2,000 m).

Tree philodendron, Filodendron
Philodendron cultivar

Family: Araceae

More than 500 species of philodendron exist in the neotropics, and a wide array are used in tropical landscaping. Various hybrids and cultivars are planted extensively as ornamentals throughout the world, doing best in warm, wet areas. This philodendron cultivar has many roots coming off of its 5- to 6-ft-long (1.5 to 2 m) stem. Its long leaf blades are heart-shaped with many fingerlike lobes along the edge. The spathe is lime green on the outside and pink inside, while the spadix is white; both are about 10 in (25 cm) long. Many of the treelike philodendrons with deeply dissected leaves are cultivars or hybrids involving the Brazilian species *Philodendron bipinnatifidum* (also called *P. selloum*).

White ginger, Lirio blanco
Hedychium coronarium

Family: Zingiberaceae

White ginger, ginger lily, butterfly lily, *lirio blanco*, and *heliotropo* are all names used for this plant, which is originally from the Himalayan region of Asia. In Costa Rica, it is a common ornamental in gardens, from sea level up to 6,600 ft (2,000 m), but it also escapes, especially into wet areas such as roadside ditches and river banks. The leaves are arranged alternately in a plane along the long stems. White ginger usually grows in clumps, 3 to 6 ft (1 to 2 m) tall. The 8-in (20 cm) flower spikes develop at the end of the leafy stems. The white flowers' heady, nocturnal fragrance attracts hawkmoths seeking nectar. When a hovering moth visits the flower and hits the reproductive structure, hairs on the blossom's anthers secrete a glue that coats the pollen as it is released, helping it stick to the insect's wings. The fruit is an orange capsule with many red seeds and red-orange arils. The rhizome from which white ginger grows looks like ginger root. Garland flower, another name for white ginger, perhaps arises from use of these flowers in Hawaiian leis.

Yesterday, today, and tomorrow
Sanjuan
Brunfelsia grandiflora

Family: Solanaceae

This large shrub is extremely attractive when in full bloom since the flowers nearly cover the whole plant with an array of violet, a softer rosy purple, and white. The blossoms fade from violet to light purple to almost white with age. Originally from South America, the plant is a popular yard plant in Costa Rica, growing in full sun at mid elevations. The genus name honors Otto Brunfels, a German botanist of the early 1500s. A number of the forty *Brunfelsia* species have important medicinal uses among indigenous peoples in South America, some of whom use *B. chiricaspi* and *B. grandiflora* in hallucinogenic concoctions such as *ayahuasca*. Studies in the 1970s led to the discovery of a substance that is not hallucinogenic but causes convulsions. It is not clear which ingredient in the plants produces the perception-altering effects.

Zamia, Yuquilla
Zamia neurophyllidia

Family: Zamiaceae

Four native species of the *Zamia* genus, one of the largest cycad genera in the world, are found in Costa Rica. *Z. neurophyllidia*, the most common of these native cycads, grows in wet Caribbean-slope forests below 3,300 ft (1,000 m) and is sometimes planted as an ornamental. It ranges from Guatemala to Peru. The zamia's roots, like those of all other cycads, contain nitrogen-fixing cyanobacteria that allow the plants to grow in poor soils. The plants are about 3 ft (1 m) tall. The leaf stalks have short spines, and there are teeth along the edge of the grooved, plastic-textured leaflets. Inch-long (2 cm) orange seeds form in cylindrical cones. *Z. neurophyllidia* is sometimes heavily defoliated by larvae of a lycaenid butterfly, *Eumaeus minyas*. This *Zamia* species was formerly referred to as *Z. skinneri*. Another of the native cycads, *Z. fairchildiana*, from the south Pacific slope, is usually tall—the trunk alone can exceed 3 ft (1 m)—and has spines on the leaf stem but not on the leaflet margins. Birds are known to disperse seeds of *Z. fairchildiana*.

Visual Index

Ageratum species • 6
Ageratum, Santa Lucía

Zantedeschia aethiopica • 7
Arum lily, Cala

Begonia species • 8
Begonias

Strelitzia reginae • 10
Bird of paradise,
Ave del paraíso

Anthurium salvinii • 11
Bird's-nest anthurium, Tabacón

Bougainvillea species and cultivars • 12
Bougainvillea, Veranera

Megaskepasma erythrochlamys • 13
Brazilian red cloak,
Pavoncillo rojo

Bromeliads, Bromelias • 14

Asclepias curassavica • 16
Butterfly weed, Viborana

Canna x *generalis* • 17
Canna lily, Platanilla

Acalypha hispida • 18
Chenille plant, Rabo de gato

Nopalea cochenillifera • 19
Cochineal, Tuna

Coffea arabica • 20
Coffee, Café

Dracaena fragrans • 22
Corn plant, Caña india

Hylocereus costaricensis • 23
Costa Rica nightblooming cactus,
Pitahaya

Chamaedorea costaricana • 24
Costa Rican bamboo palm,
Pacaya

Cheilocostus speciosus • 25
Crepe ginger, Caña agria

Codiaeum variegatum • 26
Croton, Crotón

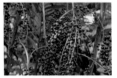

Dieffenbachia species • 27
Dumb cane, Lotería

Caesalpinia pulcherrima • 28
Dwarf poinciana, Malinche

Hamelia patens • 29
Firebush, Zorrillo real

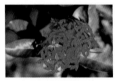

Ixora coccinea • 30
Flame of the woods,
Flor de fuego

Pyrostegia venusta • 31
Flame vine, Triquitraque

Tibouchina urvilleana • 32
Glorybush, Nazareno

Duranta erecta • 33
Golden dewdrop, Duranta

Pachystachys lutea • 34
Golden shrimp plant, Olotillo

Allamanda cathartica • 35
Golden trumpet,
Bejuco de San José

Heliconia species • 36
Heliconias, Platanillas

Hibiscus rosa-sinensis • 38
Hibiscus, Clavelón

Impatiens walleriana • 41
Impatiens, Chinas

Kohleria spicata • 42
Kohleria

Lantana camara • 43
Lantana, Cinco negritos

Lobelia laxiflora • 44
Lobelia, Caragallo

Solanum wendlandii • 45
Marriage vine, Volcán

Ipomoea species • 46
Morning glories, Churristates

Orchids, Orquídeas • 48

Carludovica rotundifolia • 50
Panama hat palm, Chidra

Carica papaya • 53
Papaya

Passiflora vitifolia • 54
Passion flower,
Granadilla del monte

Spathiphyllum friedrichsthalii • 55
Peace lily, Calita

Musa velutina • 56
Pink velvet banana,
Guineo de jardín

Euphorbia pulcherrima • 57
Poinsettia, Pastora

Stachytarpheta frantzii • 58
Porterweed, Rabo de gato

Petrea volubilis • 60
Queen's wreath, Choreque

Calathea crotalifera • 61
Rattlesnake plant, Bijagua

Cordyline fruticosa • 62
Red dracaena, Caña india

Alpinia purpurata • 63
Red ginger, Antorcha

Catharanthus roseus • 64
Rosy periwinkle, Conchita

Mimosa pudica • 65
Sensitive plant, Dormilona

Thunbergia grandiflora • 66
Sky vine, Emperatriz eugenia

Yucca guatemalensis • 67
Spineless yucca, Itabo

Monstera deliciosa • 68
Swiss cheese plant,
Costilla de Adán

Philodendron cultivar • 69
Tree philodendron, Filodendron

Hedychium coronarium • 70
White ginger, Lirio blanco

Brunfelsia grandiflora • 71
Yesterday, today, and tomorrow
Sanjuan

Zamia neurophyllidia • 72
Zamia, Yuquilla

Index